OUR AIR

G. Brian Karas

Nancy Paulsen Books

For Tina and Allie

NANCY PAULSEN BOOKS

An imprint of Penguin Random House LLC

1745 Broadway, New York, New York 10019

First published in the United States of America by Nancy Paulsen Books, an imprint of Penguin Random House LLC, 2025 • Copyright © 2025 by G. Brian Karas • Penguin Random House values and supports copyright. Copyright fuels creativity, encourages diverse voices, promotes free speech, and creates a vibrant culture. Thank you for buying an authorized edition of this book and for complying with copyright laws by not reproducing, scanning, or distributing any part of it in any form without permission. You are supporting writers and allowing Penguin Random House to continue to publish books for every reader. Please note that no part of this book may be used or reproduced in any manner for the purpose of training artificial intelligence technologies or systems. • Nancy Paulsen Books and colophon are trademarks of Penguin Random House LLC. • The Penguin colophon is a registered trademark of Penguin Books Limited. • Visit us online at PenguinRandomHouse.com. • Library of Congress Cataloging-in-Publication Data | Names: Karas, G. Brian, author. | Title: Our air / G. Brian Karas. | Description: New York, New York: Nancy Paulsen Books, 2025. | Summary: "The air plays a crucial role in our lives and in our world, making life on Earth possible, so it's important to take care of it"—Provided by publisher. | Identifiers: LCCN 2024055721 | ISBN 9780593625514 (hardcover) | ISBN 9780593625521 (ebook) | ISBN 9780593625538 (kindle edition) | Subjects: LCSH: Air—Juvenile literature. | Classification: LCC QC161.2 .K37 2025 | DDC 551.5—dc23/eng/20250108 | LC record available at https://lccn.loc.gov/2024055721 • Manufactured in China • ISBN 9780593625514 • 10 9 8 7 6 5 4 3 2 1 • TOPL • Edited by Nancy Paulsen • Art direction by Cecilia Yung • Design by Marikka Tamura and Suki Boynton • Text set in Isidora • The artwork was prepared with gouache and pencil. • The publisher does not have any control over and does not assume any responsibility for author or third-party websites or their content. • The authorized representative in the EU for product safety and compliance is Penguin Random House Ireland, Morrison Chambers, 32 Nassau Street, Dublin D02 YH68, Ireland, https://eu-contact.penguin.ie.

I am the Air.

I've been with you since you were born.
You breathe me in and breathe me out.

You don't give me much thought,
yet you know I'm always around.

So how do you even know I'm there?

You can't see me, even though
sometimes it seems like you can.

You can't smell me.
(That's not me you smell!)

And you can't hear me.
What you hear is not me
but the sound waves that travel through me.
(You can't see those either.)

But you can feel me when I move.

You can also feel what's in me. The moisture I contain has many forms. Rain, snow, ice, and fog (or lack of these things) determine what kinds of plants grow where, and how people and animals live on Earth.

I also contain little bits of matter,
some too small for your eyes to see.
They can change the color of the sky—
and sometimes make your nose itch.

The breeze coming in through your window
is one small part of a larger current,
which feeds into and becomes one of the rivers
of atmosphere that circle the planet.

**Changes in temperature and pressure affect
how I flow and blow as I move around the Earth.**

WESTERN HEMISPHERE

EASTERN HEMISPHERE

Plants and animals have learned to adapt to my many climates

and harness my energy to thrive.

What am I, exactly? I am made of a mixture of gases.

I am the perfect combination of nitrogen and oxygen for life on Earth.

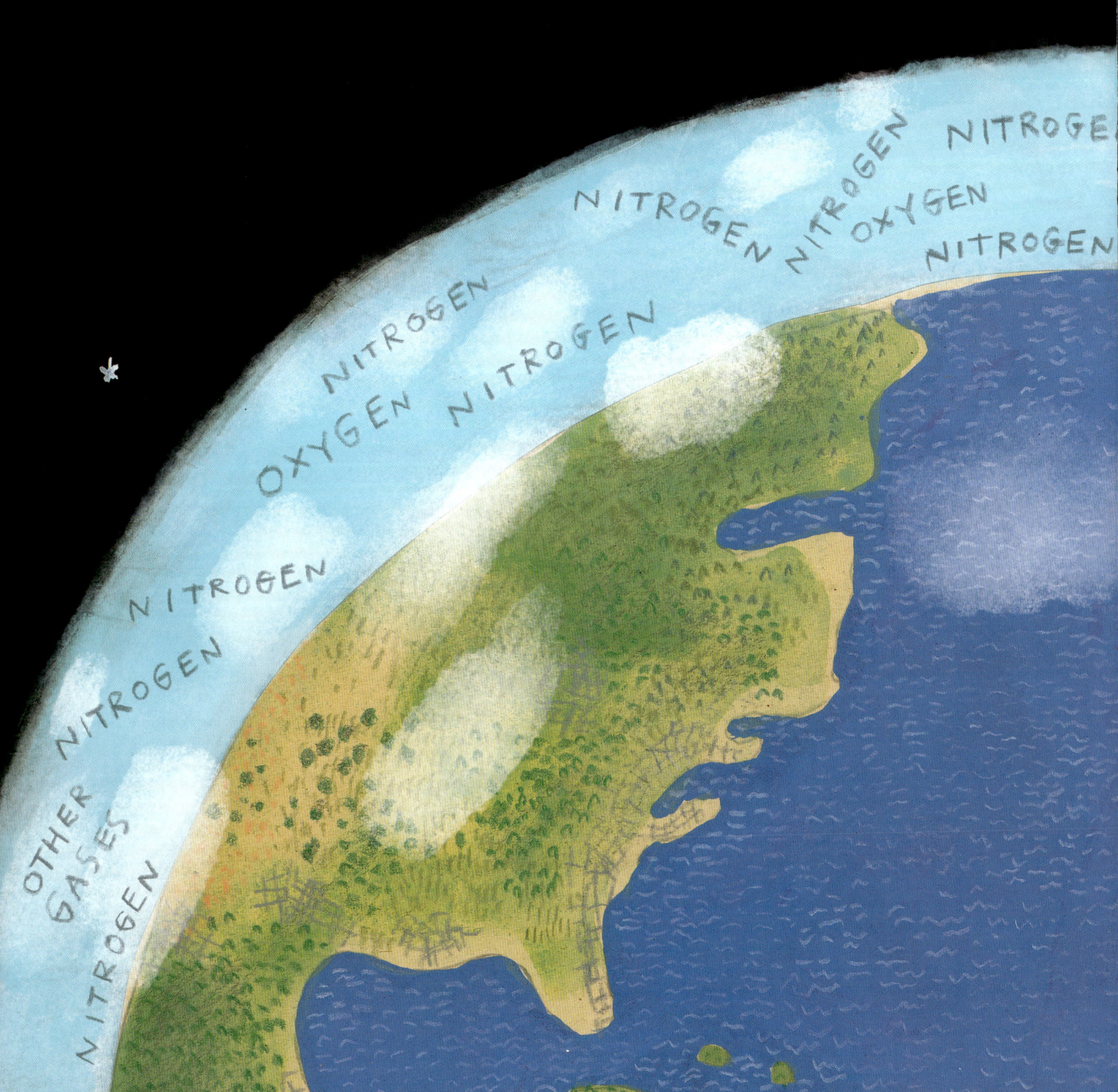

I wasn't always this way.

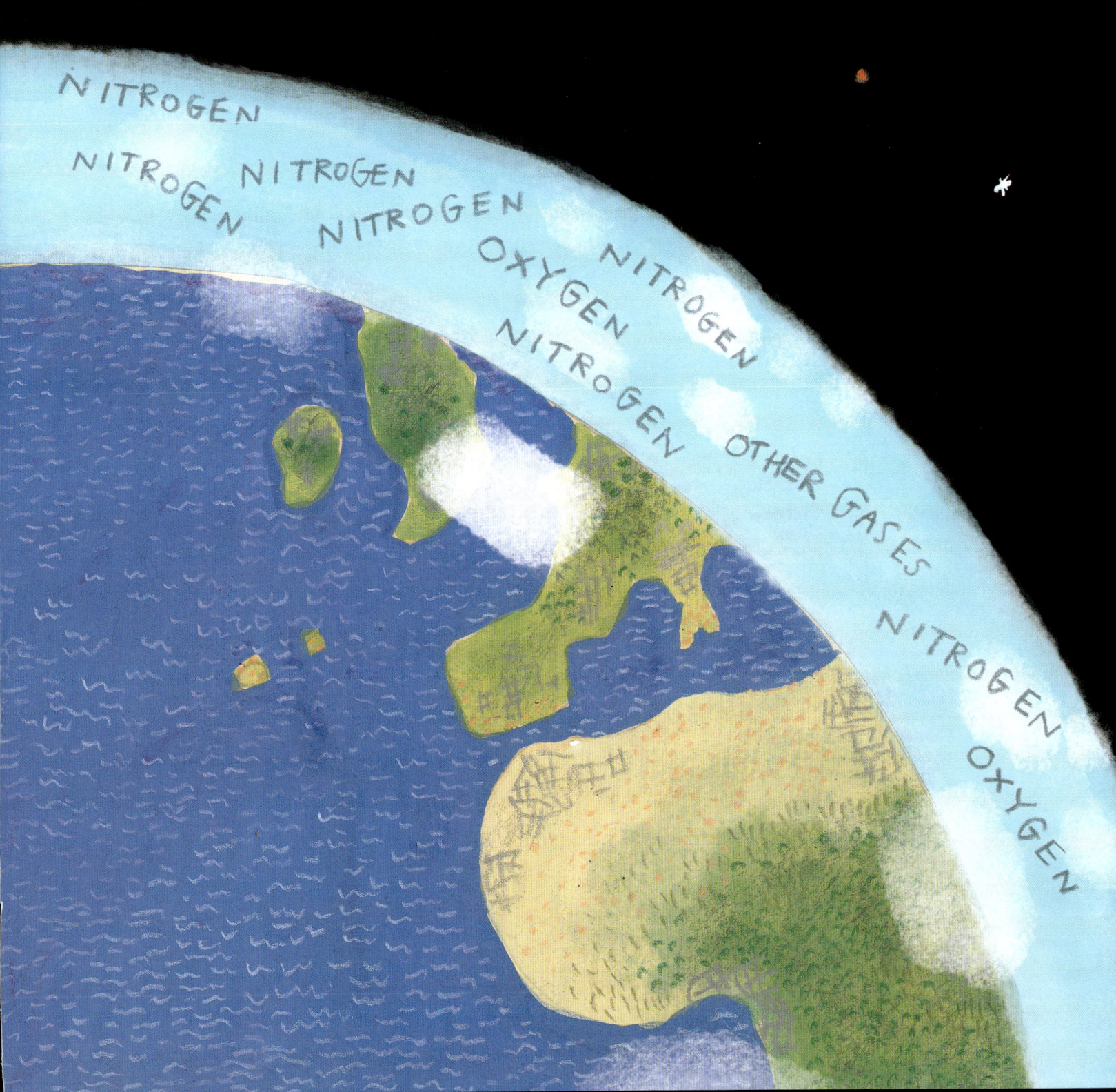

For millions and millions of years,
there were no living things on Earth.

But at this moment, you and everyone and everything are living in harmony with me.

I am your protector. I do my best to shield you from harmful rays and temperatures that are too hot or too cold. And from heavy flying things!

I'm not as strong as you may think.
There is less of me the higher you climb.
From outer space you can see
just how little of me there is.

I am not everywhere.
I'm not on every planet,
and I am nowhere in space.

You breathe me in and breathe me out.
And so does the person next to you,
and the person far away too.

Everyone is sharing me
with everyone else.

That's why it's so important to take care of me.
I am here for you
and everyone and everything on Earth.

I am the Air.

AUTHOR'S NOTE

The Earth is warmed by the Sun, more in some places than in others. These differences in warmer and cooler air, along with the Earth's rotation and land features such as mountains, plains, valleys, and bodies of water, are what create the complex circulation of air around the planet.

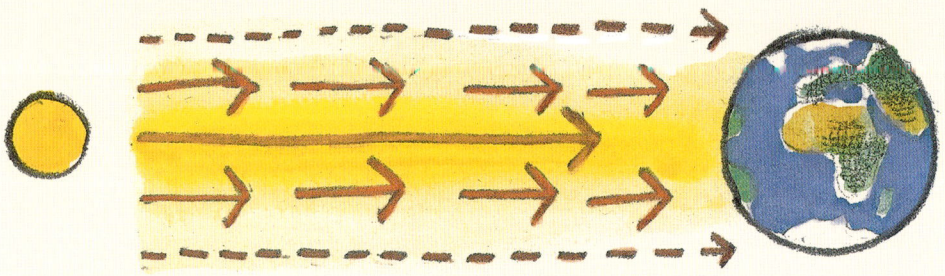

Air is densest at lower altitudes. The higher in altitude you go, the thinner the layer of air becomes.

Plants and animals exist in a symbiotic relationship and are important contributors to the chemical makeup of the air. Animals inhale oxygen and exhale carbon dioxide to live. Plants take in carbon dioxide and release oxygen. This balance allows for animals and plants to coexist and thrive.

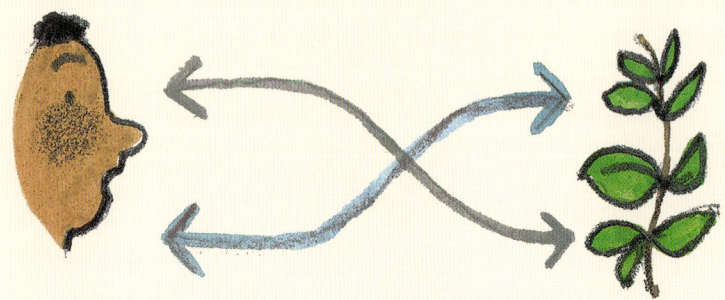